INSIDE A CONSTRUCTION MACHINE

KEITH GREAVES AND ANDREA AIRD

angus

This edition published in 2004
by Angus Books Ltd
12 Ravensbury Terrace
London SW18 4RL

ISBN 1-904594-49-2

FOR BROWN PARTWORKS
Project editor: Tom Jackson
Consultant: Dr. Donald R. Franceschetti
Designer: Sarah Williams
Illustrator: Mark Walker
Managing editor: Anne O'Daly
Picture researcher: Sean Hannaway

Production by Omnipress,
Eastbourne, UK
Printed and bound in Dubai

Contents

Machines that built history

Before mechanical construction machines were invented, heavy loads of dirt and stone were lifted, pushed, and pulled by the muscle power of people and their animals.

In the days before construction machinery roads, bridges, and buildings were built by muscle power alone. For example, large rocks were pulled on ropes along the ground on wooden sleds or rollers. Sometimes hundreds of people and animals, such as oxen and horses, were needed to pull the largest stones to where they were to be used. Dirt was carried from place to place in bags, baskets, and wheelbarrows. Digging was all done by hand with simple tools. Without construction machines it took many years to put up buildings that would take only a few weeks or months to make now.

A steam-powered tractor built in England in 1874. This machine pulled plows and other heavy loads and burned straw for fuel.

The workers at this quarry in China still have to carry stones in their hands and pull them around in simple wheelbarrows.

Simple machines

Even the most complex modern earthmoving machines are based on simple devices that were invented thousands of years ago. They are the pulley, lever, and wheel. No one knows exactly when these simple machines were invented, but by about the 1st century B.C. the Romans were using cranes powered

by people running on treadmills for lifting large blocks of stone. Modern machines work in the same way as these devices but can lift and move much more because they have powerful engines. Even the largest crane works in the same way as a simple pulley; excavators use a series of levers to pick up objects; and of course, like other vehicles, many construction machines use wheels to move across the ground efficiently.

Engine power

The most important change made to construction machines was the way that they were powered. The steam engine was invented about 300 years ago, and that meant that machines did not have to be powered by hand alone. Larger loads could be moved, and work could be done at a faster pace using steam-powered machines. Modern construction equipment now uses powerful diesel engines to produce electricity to help move heavy objects quickly and easily.

Lifting devices

Builders, engineers, and architects have always tried to figure out more energy-efficient and safer ways to lift, push, and pull heavy equipment. Although modern cranes are built from stronger material than before, their shape and structure have not changed very much for several hundred years. Most small cranes, such as the ones built into trucks, boats, or trains, have a lifting tower (called a derrick) that works in the same way as a Roman lifting machine.

GEARING UP

COG AND RUNG FOR RAISING COAL.

This machine used a simple gear for lifting coal up mine shafts about 100 years ago.

Early construction machines made use of gears. Gears use toothed wheels called cogs, which can be fitted together so that when one turns, they all turn. Depending on how they are joined together, cogs can be used to change the direction of a turning device or make a lifting machine faster or stronger.

Look inside a construction machine

Construction machines are designed to do many different jobs, and so they often look different from each other. However, they all share many things that help them do their job.

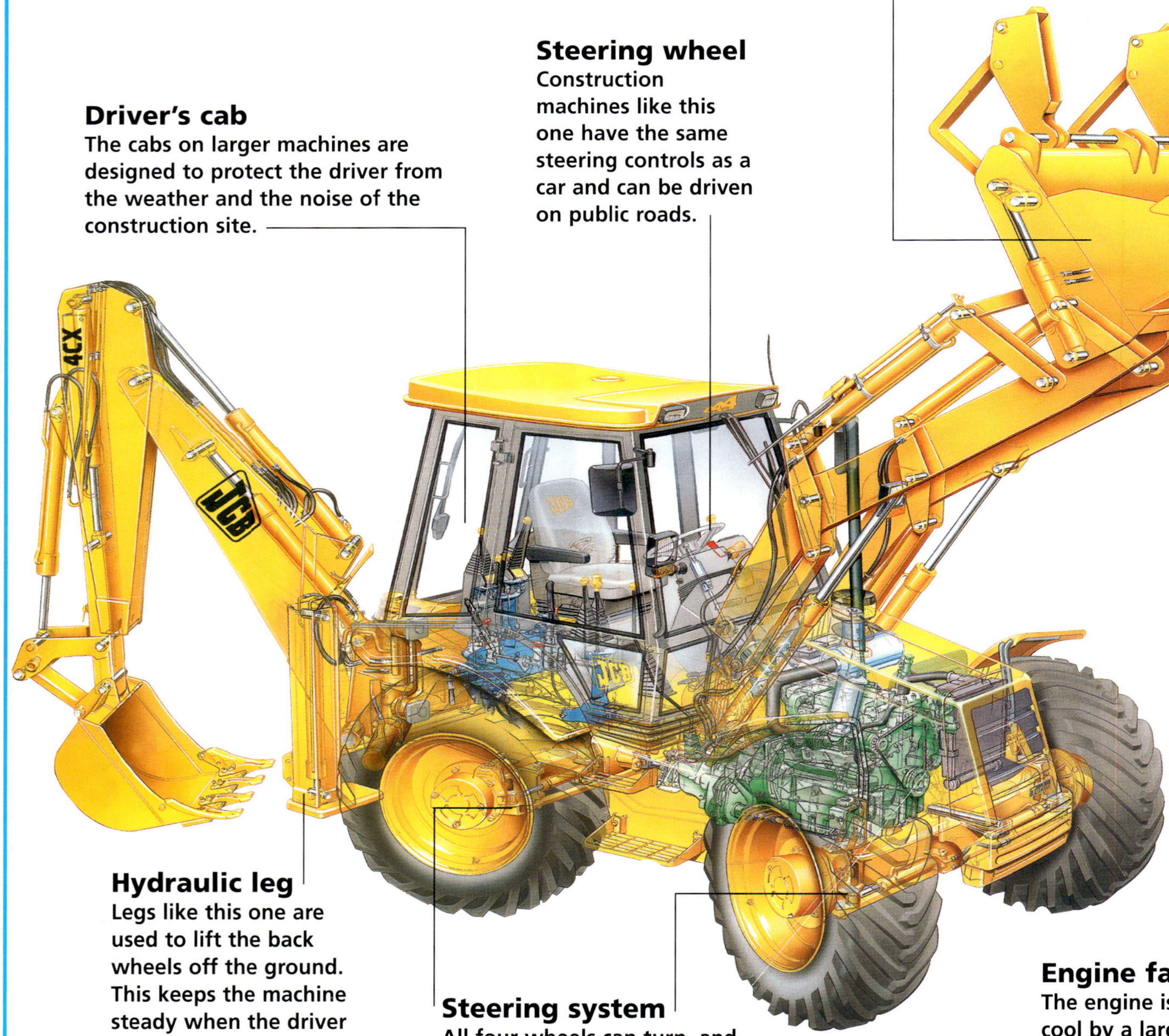

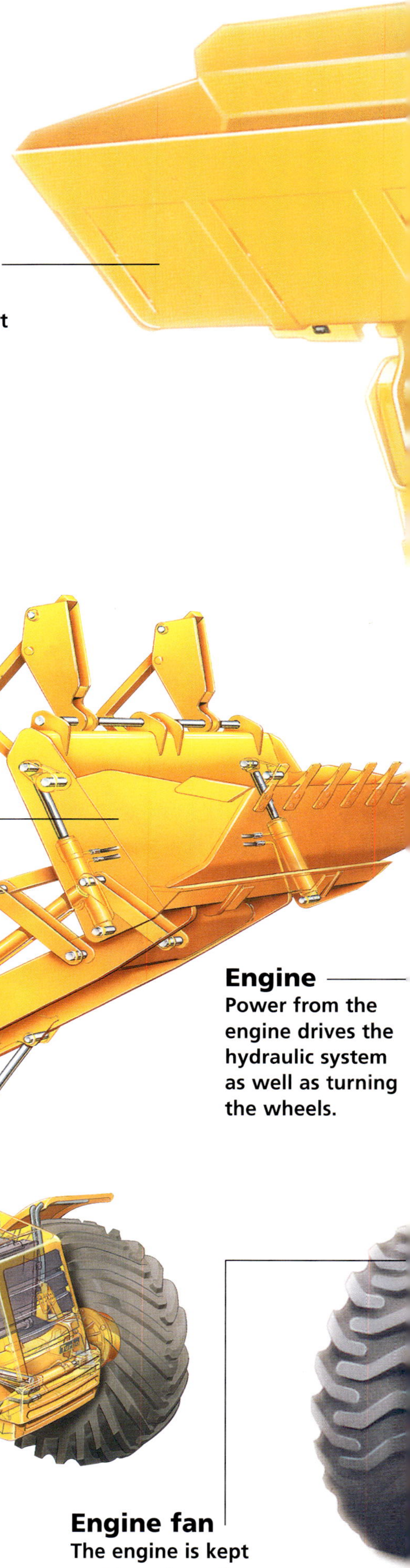

Backhoe loaders

Hydraulic hose
The fluid used to move the arms and buckets is pumped through hoses like this one.

Driver's chair
The driver's chair can turn around so the driver can see clearly when using the backhoe.

Hydraulic piston
Fluid from the hydraulic hoses pushes on pistons like this one to move the machine's arms and buckets.

Controls
The driver can steer the machine and control the loader and backhoe using these levers and joysticks.

Heavy tyres
The big treads on these tyres help the machine drive through mud and sand.

Chain transmission
Power from the engine is sent to the wheels by strong chains.

Backhoe bucket
This bucket is very manoeuvrable. It is used to dig holes and trenches.

Using engine power

Modern construction machines use large diesel engines. The engine drives the machine from place to place, as well as moving its lifting, pulling, pushing, and digging equipment.

Diesel fuel is similar to petrol. Large machines use diesel engines because they are more efficient than petrol ones. A diesel engine sucks in air and squashes it together, making the air very hot. When fuel is added to this compressed air, the mixture burns immediately, and the force of the explosion is used to drive the engine.

An earthmover's engine is not only used to move the machine around, it also works many different components on the machine. For example, a single tractor engine can be used to turn the wheels, control digging machinery, heat the driver's cab, turn on the lights, and operate the windscreen wipers.

Power transfer

Power from the engine is transferred, through a device called a clutch, to long metal rods, called axles, that connect the wheels of the

Some construction machines have huge engines, like this large diesel. Diesel engines are very efficient for large machines.

LIFT A WEIGHT WITH AIR

You will need a balloon and some heavy books.

1 Put the books on top of the balloon close to the edge of the table.

2 Blow air into the balloon using a pump.

3 The books rise up. This is how hydraulics work.

vehicle together. The clutch can be worked by the driver using a foot pedal, or it can run automatically. The engine's power then makes the wheels turn around.

Some engines are attached to moving parts by chains, just as a bicycle wheel is attached to the pedals. Chains are better at controlling the movement of machinery that often needs to go in different directions.

Hydraulic arms

A hydraulic system uses a liquid that cannot be squashed, such as oil, to move energy from one place to another. The liquid is pumped through the system by motors. Hydraulics can make a small force into a larger force by reducing the size of pipe that the liquid is pushed through. Hydraulic machines need to be "bled." This is a process that removes any air bubbles from the liquid. Without it hydraulic systems would soon stop working.

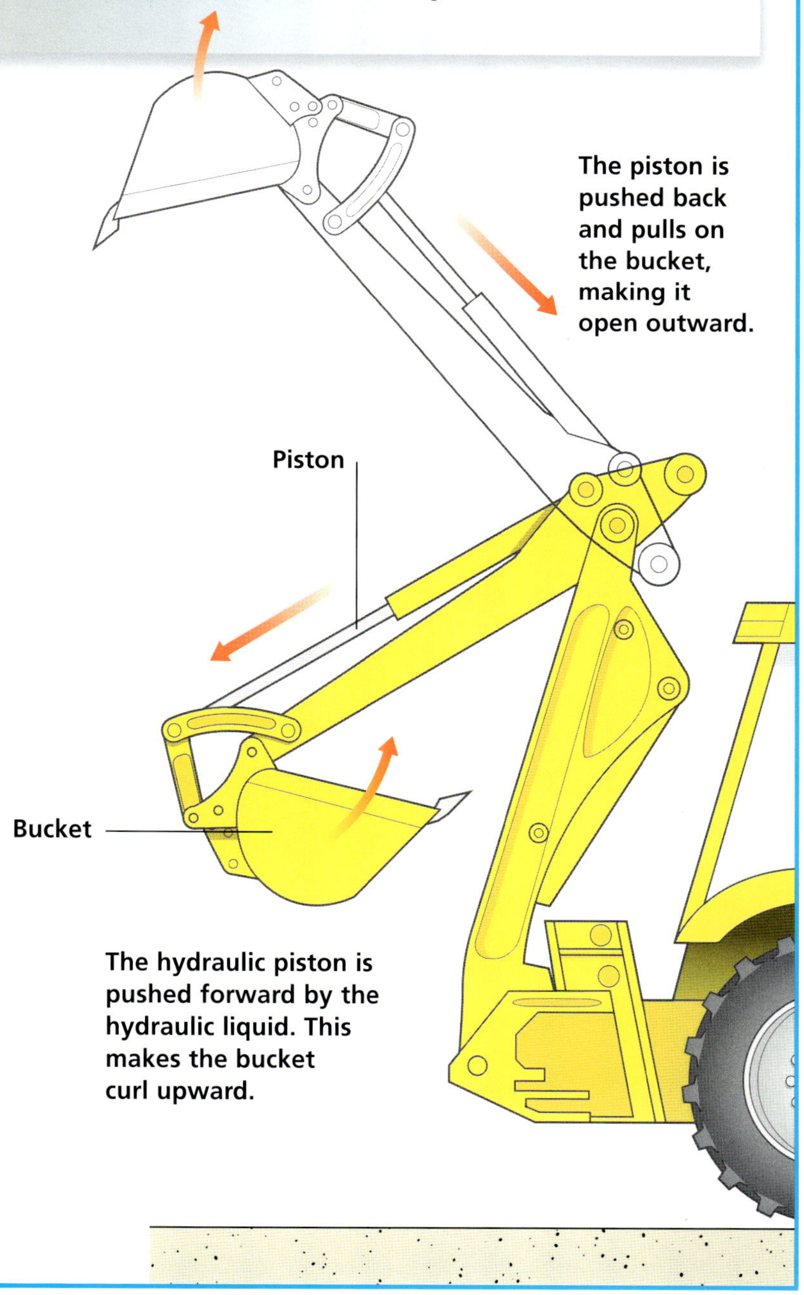

Right: This diagram shows the movement of an excavator's bucket. Hydraulic liquid pushes on pistons to lift or lower the arms and bucket.

Making tracks

Construction machines are very heavy and need special types of tyres and wheels so they can be steered easily over rough ground.

The tyres on construction machines are usually very wide and tall. This is so they can support the weight of the machine and its heavy load. Wide tyres exert less pressure on the ground than thin tyres. The weight is spread out over the wide tyres. This keeps the machine from sinking into soft ground.

Some earthmovers have caterpillar tracks. Caterpillar tracks are used when the ground is very rough and normal tires could puncture easily. Construction machines with tracks are better at travelling up and down steep hills, and they can drive across wide ditches without getting stuck as wheeled machines would.

Steering

Some construction machines operate by steering the two front wheels. Other machines use a four-wheel-steering system that turns all

The tyres from some machines are very large indeed. This truck tyre is taller than the driver! The deep tread helps the tyre grip the ground.

four wheels at once, so the driver can turn the machine completely around in small spaces. Four-wheel-drive vehicles have all their wheels connected to the engine instead of just two. This helps them grip on loose surfaces.

Some long trucks are articulated—in two parts that are connected by a hinge joint. This joint system makes the vehicle more flexible so it is easier to turn.

Tracked machines have a different steering system entirely. The controls used to steer a tracked vehicle look very different from a normal steering wheel. Two levers in the cab control the power and direction of each track. To turn the machine, the driver reduces the power to one track so it turns slower than the other, and the vehicle swings around. With this steering system tracked vehicles can spin around on the spot.

This tracked excavator can get to places wheeled machines cannot.

MOVING WEIGHTS WITH ROLLERS

You will need a shoebox, a rubber band, some round pencils, and a few stones.

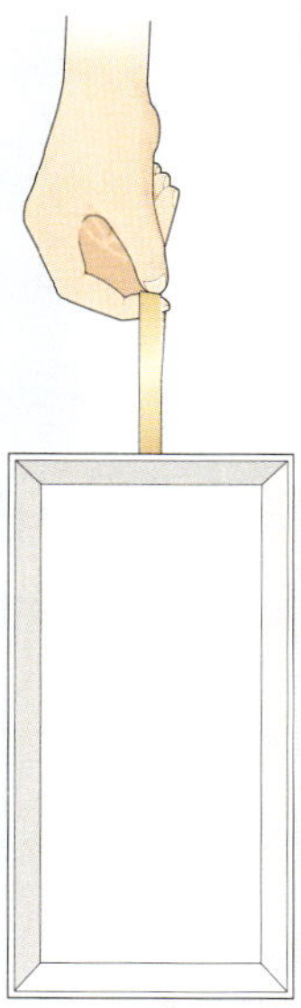

1 **Attach the rubber band to a shoebox. Pull the box along the ground. Note how much the band stretches.**

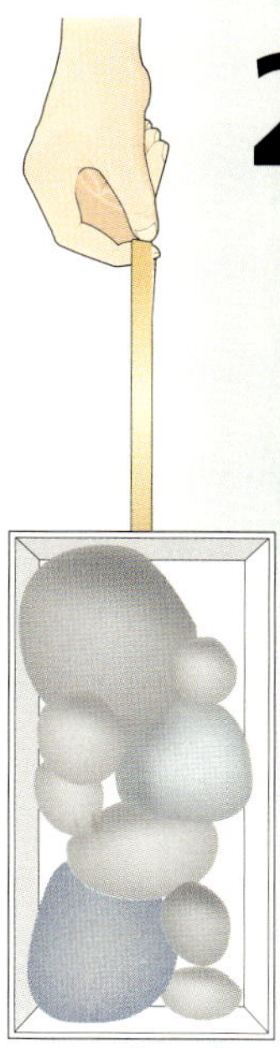

2 **Fill the box with a few stones. How much does the band stretch now?**

3 **Now put the box on round pencils or pens. Does the band stretch as much as before?**

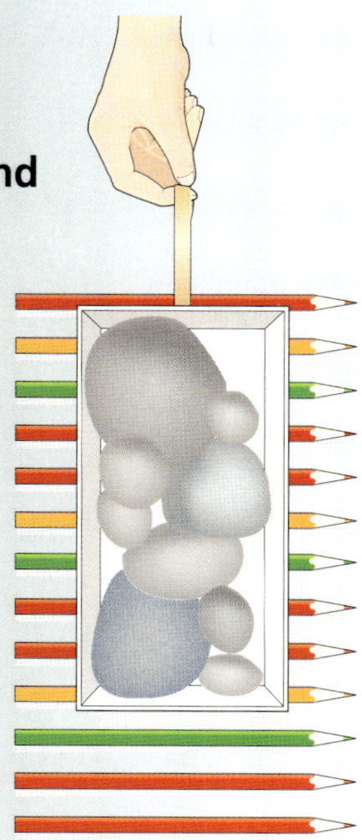

The stones make the box heavier and increase the friction force between the box and the ground. The pencils can roll, and that reduces the friction so the heavy box can move more easily.

Preparing the ground

Building roads is a big job. Several types of construction machine are used to turn a rough piece of land into a smooth road.

The first thing road builders do is remove all of the unwanted stones and tree stumps. Heavy mechanical claws, called rippers, are attached to the back end of a tractor or dozer and used to break up hard stones and pull tree stumps from the ground.

Leveling the ground

A machine called a scraper levels uneven ground. Scrapers have two main parts: a scraper bowl and scraping blades. When the scraper moves forward, the scraper blades rotate, like a moving escalator. As the blades move, they pick up earth and drop it into the scraper bowl. Scraper bowls are at the back of the machines and can hold a load of about 17 tons (16.5 tonnes).

Grading is the process of making rough ground level. Grader machines act like large sieves, removing larger pieces of rubble and leaving smaller pieces behind. This construction machine drags a large blade across the ground behind it. The blade collects larger stones and bits of rubble, and

This machine is a grader. The large blade underneath it is dragged along the ground, pushing loose dirt and stones to one side.

A roller being used to level a new road, which is being built in Russia. The roller's heavy wheels squeeze the road surface flat.

pushes them to the sides of the road, sometimes into a ditch. Graders can also be used to dig trenches and ditches by angling the blade into the ground.

Road rollers

If a road has not been made to be strong enough, it will soon break up after heavy traffic starts driving on it. To stop this from happening, the different layers of a road must be flattened until they are hard and smooth. The machines used for this job are called rollers. The weight of the machine squeezes the layers together. Large rollers are driven backward and forward over the new road surface, simply flattening and smoothing the surface with their heavy wheels. Some rollers can weigh almost 7 tons (6.8 tonnes)!

ADDING THE TOPPING

A road-paving machine laying asphalt. These machines move very slowly.

The top layer of many roads is made of asphalt, which is a mixture of small stones and tar (a very sticky oily substance). Road pavers are machines that are designed to lay an even surface of asphalt that is free of any lumps and bumps. The machine spreads a layer of hot, soft asphalt on the road. When the asphalt cools down, it forms a hard and flat surface that is very strong.

Mobile lifters

Mobile cranes have the advantage of being able to move around from site to site. Mobile cranes can be mounted on trucks, caterpillar tracks, boats or even trains.

The lifting arm, or jib, of a mobile crane must be able to fold up to a small size so the crane can be moved easily and safely. Trucks that carry cranes have large tires to support the weight of the crane and are designed to be driven along normal roads. For road journeys the jib is lowered and its sections folded inside one another. The pulley block is clamped onto the front of the truck. This keeps the crane steady while the truck is on the move. The biggest truck cranes can lift loads of 100 tons (98 tonnes) and have jibs that can extend to 450 feet (135m).

Cranes with long jib arms are often only able to lift small weights compared to cranes with shorter jibs. This is because longer jibs break more easily, and they also make the crane less stable.

A truck crane lifts a wind turbine up to the top of a mast at a wind farm in California.

Keeping stable

Truck cranes have four metal legs called outriggers that lift the truck off the ground when the crane is being used. Outriggers help keep the crane steady and also take the weight off the wheel to prevent damage to the truck's suspension. Obviously, truck cranes cannot move when they are lifting. However, caterpillar cranes are more stable than truck cranes, and so they can move even while they are carrying a heavy load.

Once the truck has been secured, the jib can be raised. Some truck cranes have jibs with sections inside each other that extend like a telescope to make the jib taller.

A crane with caterpillar tracks at the South Pole. The tracks help the crane move over the ice.

PULLEY POWER

You will need two broomsticks, a long rope, and four friends

1 Ask your friends to pull against you on a rope. Who wins? They do, because together they are four times as strong as you are.

2 Now wrap the rope around two broomsticks as in the diagram and ask your friends to hold each broomstick. When you pull against them now, who wins? You do because you are using a pulley system. Pulley systems are used by cranes to help them lift heavy weights.

Cranes at work

Cranes come in different shapes and sizes for different types of jobs. Cranes in automobile and aeroplane factories are attached to roof beams so they take up less space in the factory and because it would be too dangerous to have them driving around on the floor.

There are even cranes built on boats. These floating cranes are used to build bridges or structures like oil platforms. The largest floating cranes can hold loads of 2,700 tons (2,646 tonnes). That is about the same as lifting 120 trucks!

Tower crane

Tower cranes are tall machines that are used to lift and move heavy loads. They are the tallest cranes in the world, but unlike mobile cranes, they are fixed in one place.

Tower cranes are made from a series of strengthened squares that are bolted together to make a crisscrossed structure. This structure is very strong because the crane's weight is evenly distributed. The crane has no flat surfaces to catch the wind, which would rock the crane dangerously.

Large blocks of concrete called counterweights are placed at the far end of the crane's jib to keep the tower crane stable. If they did not have counterweights, cranes would fall over when they lifted heavy loads.

Building a tower crane

Tower cranes are built section by section on building sites. The base is secured to the ground by bolts or heavy weights. A smaller crane is then lifted onto the tower frame. This crane builds up the tower by adding more and more sections around itself, making the crane taller. When the tower is tall enough, the jib, the driver's cab, and the counterweights are added.

Pulleys

Pulleys make things easier to lift. A pulley is a grooved wheel around which a cable or rope is passed. Pulling down on the other end of

Crane operators sit in a cab high up at the top of the crane. They talk to people on the ground using a radio.

Tower cranes are very tall structures, but they are only temporary.

the cable lifts the load. It is much easier to pull on a cable than it is to lift a load up. If the pulley uses more than one wheel, it makes it even better at lifting weights. For example, it is four times easier to lift a weight using a pulley with two wheels than it is when using a pulley with just one. However, the cable must be pulled four times as far.

Skyscrapers are built using cranes that have been put up on top of the building itself. Many tall buildings have a crane on their roofs to hold a window cleaners' platform.

Right: A diagram of a tower crane. Cranes like this are very complicated and need very skillful operators to work them.

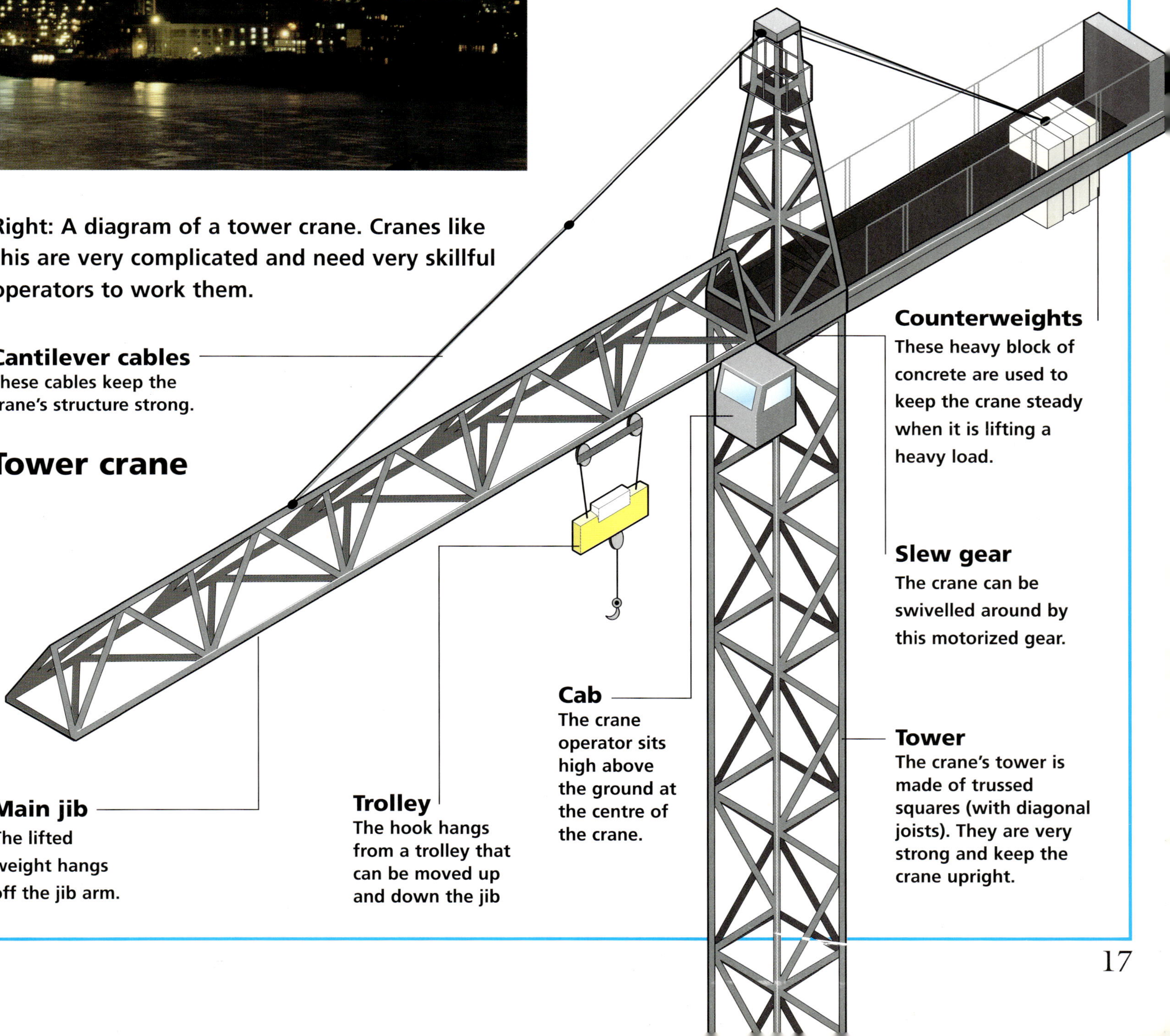

Excavators–expert hole diggers

Excavators are digging machines. They have powerful arms that scoop up and lift out large quantities of dirt and rock.

Excavators come in many types and sizes. Smaller machines are used for ditch digging and ponds; larger excavators can hold over 20 tons (19.5 tonnes) of rock and soil at once! Arm-and-bucket excavators have shovel-like buckets with large, tough steel claws that bite into the ground. The bucket is on the end of a movable dipper arm. The arm is in turn joined to a boom, which is attached to the front of the machine. However, even the biggest arm-and-bucket excavators can dig to only about 15 feet (4.5m).

Backhoe loaders have attachments at the front and back that make them extremely useful machines. The attachment at the back of the machine is called a backhoe. A backhoe is a small version of an arm-and-bucket excavator and is often used to dig small holes. A shovel-like loader is attached to arms at the front of the machine, which can be used to lift up dirt and rock.

A huge dragline excavator at an open-pit coal mine in Ohio. Its bucket can hold two buses!

TYPES OF EXCAVATOR

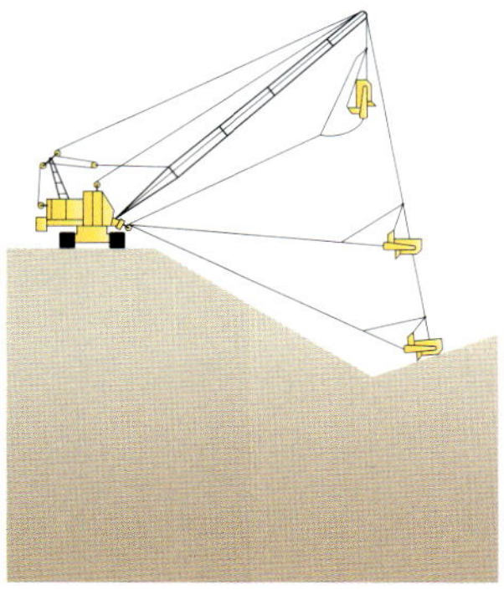

Above: Dragline excavators scrape up rock and earth from the ground using a huge bucket.

Below: Arm-and-bucket excavators are used to cut away at cliff and hill faces to get at metal ores, coal, and other valuable rocks.

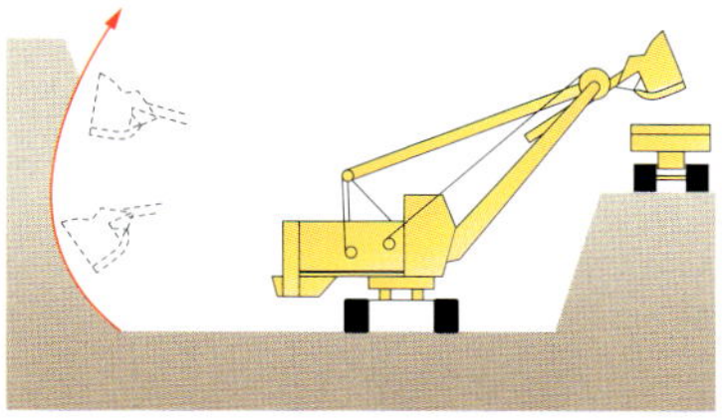

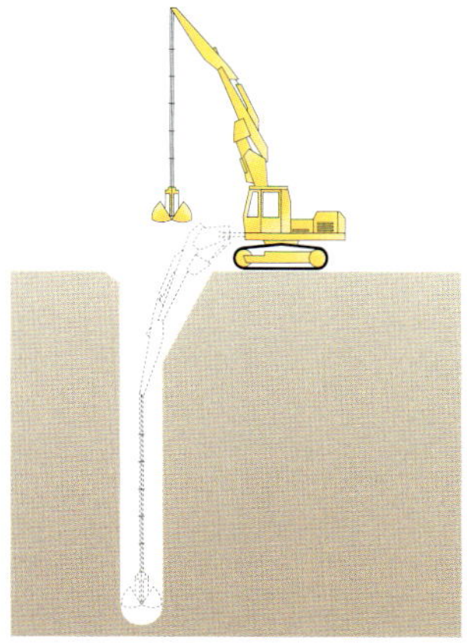

Above: Clamshell grab excavators can dig very deep but narrow holes suitable for wells.

Below: Backhoe and front-mounted excavators have a bucket on a flexible arm that can dig out holes and clear piles of earth.

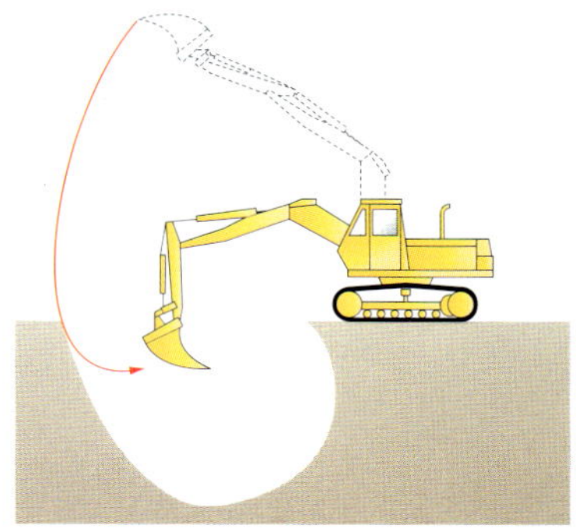

Drop and drag

A dragline is used to dig along slopes. Instead of an arm, the bucket is hung from a couple of cables. One cable lowers the bucket from a long boom until it hits the ground. Then the bucket is pulled up the slope by the other cable. As the bucket is dragged across the ground, its steel teeth rip into the ground, and it picks up large loads of soil and rock.

When deep narrow holes need digging, a clamshell excavator is used. This machine's bucket is called a clamshell because it has two parts that can open and close like jaws or the shell of a clam. The clamshell is lowered on a cable into the hole. When it gets to the bottom of the hole, the clamshell bites a chunk of the ground and then is pulled up to the surface.

Excavators at work

You may have seen backhoe excavators dig large holes for human-made lakes. But these digging machines can be used for all sorts of jobs. Smaller excavators have been used in archaeological digs, dredging up shipwrecks, and even digging graves!

An arm-and-bucket excavator fills a truck with a load of rock and earth.

Tractor, loader, dump truck and bulldozer

Construction machines are designed to do just one or two jobs really well, such as pulling, pushing, or lifting. This means that different machines often have to work together on a building site.

A tractor is a self-powered work machine. It pushes or pulls a variety of other machines over land. Tractors are used in agriculture, mining, and road making. There are wheeled tractors, which have large tyres, and crawlers, which move on heavy tracks. A tractor can plough, grade, cut, and can be used to pull cranes. Larger tractors can control attachments such as bulldozer blades, rippers, or snowploughs.

Most earth moving is done by machines that simply push and shove. A bulldozer is one such machine. The huge steel blade at the front of a bulldozer shifts loads by pushing them out of the way into big piles. The blade can be tilted and moved up and down by lifting arms. That way the driver can control where the load is pushed.

Loaders and dumpers

Machines designed to lift and carry soil and rock are called front-end loaders. They have shovel-like buckets at the front. The metal bucket scoops up piles of earth and can be tilted up and down by supporting arms. The dirt is loaded into dump trucks to be carried away.

A front-end loader carries a large load of earth over to a dump truck. These two machines often work together on construction sites.

As well as being used to power and pull many types of construction machine, tractors are commonly used on farms.

Dump trucks move the unwanted soil and rock away from the construction site. The truck body is V-shaped so that when the truck is loaded, most of the weight is in the center. This makes the dump truck stable. When the load is ready to be dumped, the body is raised by hydraulic rams, and the dirt and rock fall out an open gate at the back of the truck. Large dump trucks can carry a load of over 90 tons (88 tonnes) of dirt, which can slide out in a just few seconds.

FACT FILE

- **Dump trucks have very flexible suspension systems that help them travel over unlevel ground without dropping their heavy load.**

- **Bulldozers are used by armies to clear ground quickly and make defensive earthworks in war zones.**

MOVING PILES OF EARTH

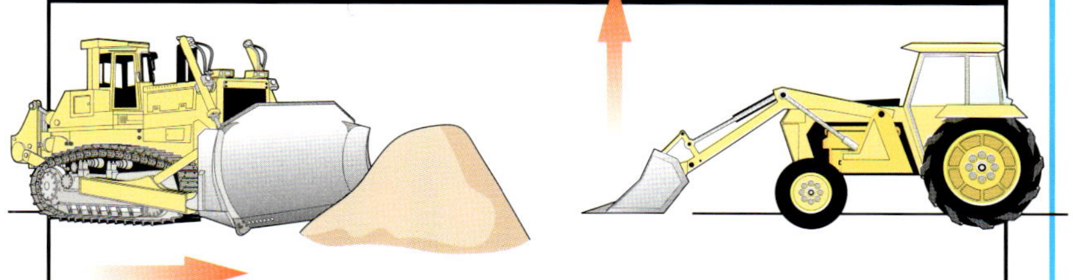

A bulldozer pushes dirt into a pile, and a loader scoops some of it up in its bucket.

The loader drops the earth into a dump truck. The truck can hold many tons of dirt.

When the truck is full, it is driven to another site where it dumps the dirt.

Unusual uses for machines

Construction machines are not only found on building sites. Some machines are designed to do special jobs, and others are used in unusual places.

Above: A drilling machine is used to bore a long hole into the ground in a surface mine. This drilling machine can also be used to take samples of rock from different depths under the ground.

Right: A loader fills a small dump truck with excavated rocks from a gold mine in Zimbabwe, Africa. The rock is taken up to the surface, where the gold is extracted (removed) from the rock.

Construction machines can turn up in unlikely locations, doing uncommon things. Machines that are very similar to those found on normal construction sites, such as loaders, excavators, and dump trucks, are used in mines where rocks and dirt are extracted for metals, such as gold and iron, and stone, such as limestone and gravel.

Some mines are on the surface, but many of them are underground, and the machines are taken down to work below the surface. Large machines are taken down by lifts in pieces and put back together below the ground. Other

machines are used in the mining industry to drill holes. Holes are needed for several reasons, such as to lay explosive charges or to find out if there are valuable substances underground in that area.

Excavators called dredgers do not operate under or even on the ground. They float on water and are designed to excavate the bottom of the sea or rivers. Dredgers are commonly used to clear shipping routes of sand and silt. If the routes are not dredged, large boats would run aground on the seabed. Dredgers are also used to dig holes for the foundations of bridges and other buildings that cross over stretches of water.

Right: A large dredger works to keep a shipping channel in Hong Kong, China, open to traffic. The huge clamshell picks up dirt from the seabed and drops it into the barge beside the dredger.

Giant machines

Some earthmovers are enormous. The world's largest machines are used in mining and quarries to excavate and haul thousands of tons of earth and rock in a single day.

In large mines it is cheaper to use a few very large machines than many small ones. The buckets of the largest excavators hold around 800 tons (784 tonnes). In some cases several buckets of this size are attached to a huge movable wheel. The machine is driven on caterpillar tracks up against a slope, and as the wheel turns, the buckets scoop up many tons of rock and drop it onto a conveyor belt. The conveyor carries the rock to be hauled away on trains, trucks, or boats.

The dump trucks that are used to take away unwanted earth and rock from the work site are huge as well. A mining truck can hold up to 150 tons (147 tonnes), which is the equivalent of about 30 elephants. When the truck is ready to dump a load, the driver operates the controls from the cab. The tipper body is pushed upright by massive hydraulic

A huge excavator at an iron-ore mine in Australia. The excavator and its conveyor both have tracks so they can be moved around the mine.

This massive ore-hauling truck has a computer-controlled steering and suspension system, making it easier and safer to drive.

arms so the load slides out. To move this enormous weight, mining dump trucks must have extremely powerful engines. Some have engines that are as powerful as 1,500 horses.

On the move

These machines are too big to travel on ordinary roads. When the giants are moved from site to site, they must be taken apart and reassembled on the new site. Once assembled, special wide roads are built that enable the machines to move around.

Smart machines

Machines used in mining haul many tons of dirt and rock in a single load. The weight of the machine can increase very quickly as it is loaded with rocks. This causes an incredible amount of strain on the wheels and suspension system. It is important for the machine to be able to adjust the suspension quickly and easily. Today's gigantic machines are equipped with computers that control and adjust the suspension depending on how much load the machine is carrying.

FACT FILE

- **Some giant excavators have been given names such as Big Hog, The Silver Spade, River King, and Big Brutus. When they have finished their work, many of them are turned into tourist attractions.**

- **Giant dragline machines take about two years to build but can scoop up over 1,500,000 tons (1,470,000 tonnes) of coal or metal ore in a year.**

Below: A worker is dwarfed by a huge excavator wheel. To save money, many of these machines work all day and night.

NASA's rocket mover

NASA's enormous rockets and space shuttles have special machines to move them from the place where they are put together to their launch site.

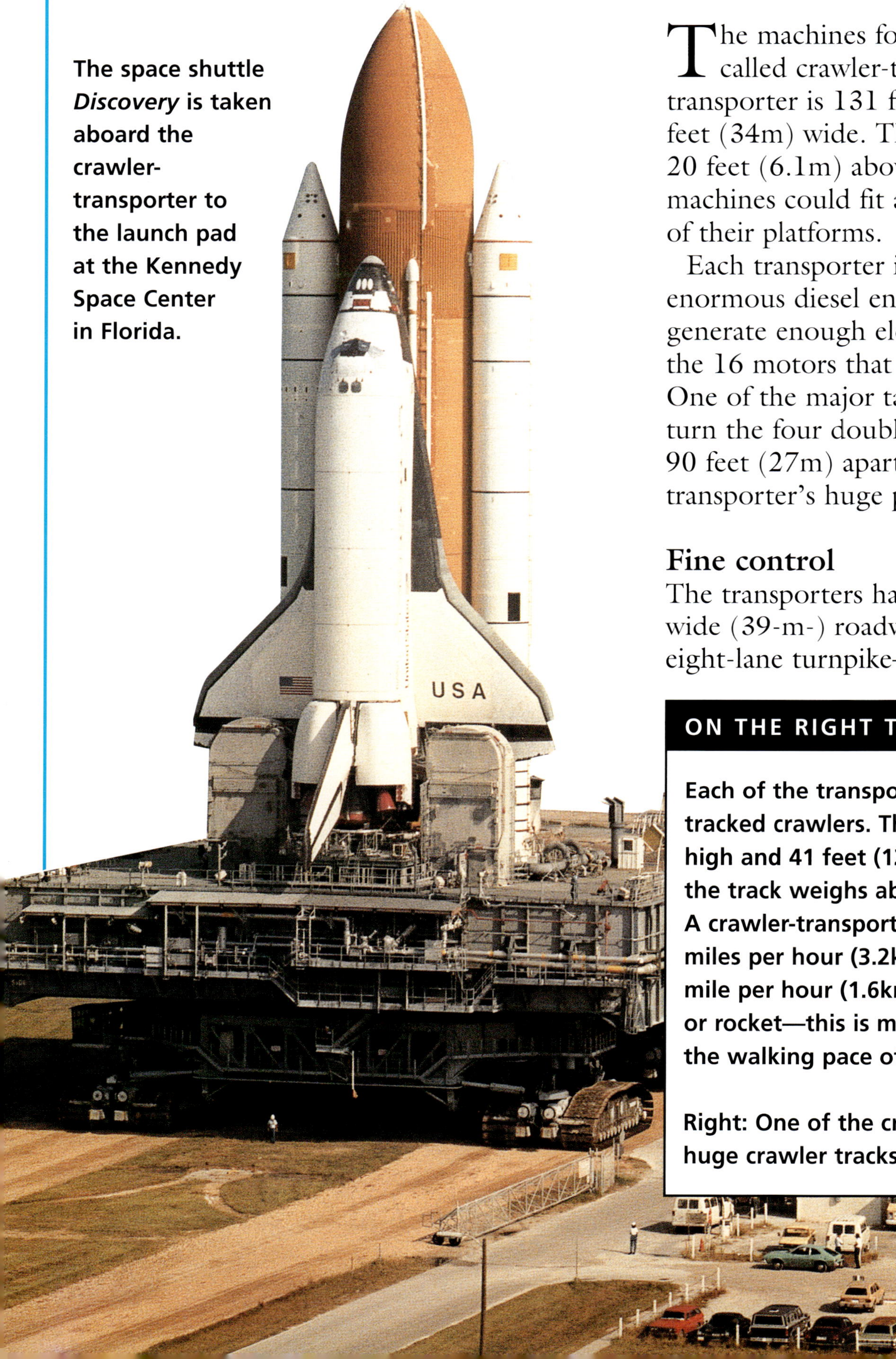

The space shuttle *Discovery* is taken aboard the crawler-transporter to the launch pad at the Kennedy Space Center in Florida.

The machines for moving spacecraft are called crawler-transporters. Each transporter is 131 feet (39m) long and 114 feet (34m) wide. The transporter's platform is 20 feet (6.1m) above ground level. These machines could fit a baseball diamond on top of their platforms.

Each transporter is powered by two enormous diesel engines. The engines generate enough electrical power to operate the 16 motors that perform a variety of jobs. One of the major tasks for the motors is to turn the four double-tracked crawlers spaced 90 feet (27m) apart at each corner of the transporter's huge platform.

Fine control

The transporters have to travel on a 130-feet-wide (39-m-) roadway—almost as wide as an eight-lane turnpike—that joins the rocket-

ON THE RIGHT TRACK

Each of the transporters moves on four double-tracked crawlers. These tracks are 10 feet (3m) high and 41 feet (12m) long. A single link on the track weighs about 2,000 pounds (906kg). A crawler-transporter's maximum speed is two miles per hour (3.2km/h) when unloaded or one mile per hour (1.6km/h) when carrying a shuttle or rocket—this is much slower than the walking pace of a person.

Right: One of the crawler-transporter's four huge crawler tracks.

FACT FILE

- NASA scientists have calculated that their two crawler-transporters have travelled 2,526 miles (4,041km) since they were built in the 1960s. That is the same as driving from the space center to Los Angeles, California.

- The two huge vehicles were originally built to carry the huge Saturn V space rockets designed to fly to the Moon. Now they are used to carry NASA's four space shuttles.

Above: A Saturn V rocket about fly to the Moon in 1969. A crawler-transporter had delivered the spacecraft to the pad and lowered it into place.

assembly buildings to the complex of launch pads at NASA's (National Aeronautics Space Administration) Kennedy Space Center in Florida. This road goes up a very slight ramp leading to the launch pad, and the transporters have a levelling system designed to keep the top of the rocket or space shuttle vehicle steady and stop it from falling over. Hydraulic arms are used to keep the transporter's platform flat all the time.

When a space vehicle is ready to be taken to the launch pad, a crawler-transporter is driven into position beneath it. The spacecraft is already on its launcher platform. The transporter picks up this platform and rolls to the launch pad. It takes five hours to make the trip at a top speed of about one mile per hour (1.6km/h). The crawler transporter is not very fuel-efficient. It burns about 150 gallons (568l) of diesel every mile. Cars use about about a thirtieth of a gallon per mile!

Future construction

Engineers and scientists face challenges all the time when trying to design new machines for working in hostile environments.

A satellite is launched into orbit from a space shuttle. Objects are weightless in space. On Earth this satellite would have to be moved by a crane.

Scientists are already developing machines to mine materials from other planets, the Moon, and even asteroids! Planets and asteroids contain many precious materials such as silver and gold, but the biggest problems will be trying to land machines on them and then getting the rocks back to Earth.

International Space Station

The different sections of the International Space Station are very difficult to move on Earth because of their bulky design, but in orbit around the Earth they behave as if they are weightless because everything is falling around the world at the same speed. This means a single astronaut can move an object that would have to be moved by crane on Earth. The International Space Station will be the biggest human-made object in space to date when it is completed.

A space-station module is lifted by a crane at the factory in which it was built. In space a crane will not be needed to move the module.

Ocean explorers

Most of the Earth is covered in seawater. There are huge amounts of natural resources, such as oil, hidden deep beneath the water. In deep water one of the only ways to find precious sites is to use submersible vehicles that can crawl along the seabed and take samples and photos of the terrain. The pressure caused by the deep water is so great that it would be very difficult to put people in them, so these machines are controlled by computers from ships floating on the surface.

FACT FILE

- In the future, construction machines could be controlled using virtual reality. This would allow machines to be used in places where people can not easily get to.
- Think of different ways that earthmovers and construction vehicles could be used in the future. Draw a picture of your new design.

Underwater mining machines could be used to collect rocks and excavate holes in the seabed deep below the water's surface.

Airborne cranes

Cable cars are built on mountains to help people get to and from the summit easily. Cable cars need huge towers and heavy cables to run on. The only way of getting the materials to these places is by air. Huge airships or powerful helicopters that can hover over the construction site are already used to move the construction equipment, such as excavators and loaders, to remote places. They are also used as cranes to lift objects around the construction site.

UNDERWATER MINING

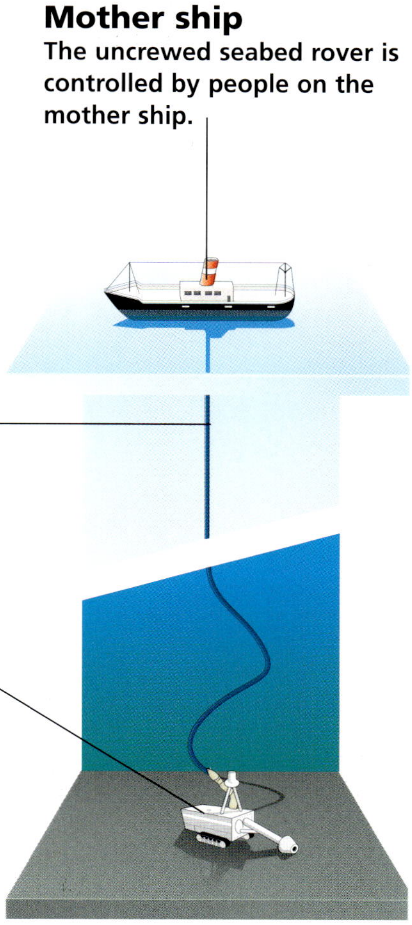

Valuable minerals could be collected from the bottom of the ocean using an uncrewed rover that would travel across the seabed.

Glossary

ASPHALT—a mixture of stones and oily tar that is used to cover road surfaces.

AXLE—the long metal rod that joins two wheels together.

CANTILEVER—a support for a horizontal structure like a floor or jib arm.

COG—a toothed wheel used in gears. One cog's teeth fit into another's.

CONVEYOR—a machine that uses a large moving belt to transport objects.

DIESEL OIL—a type of fuel used in the engines of large machines.

ESCALATOR—a mechanical stairway.

EXTRACT—to take out or remove a substance from rocks or dirt.

FRONT-END LOADER—a machine that has a loader bucket attached to arms sticking out the front.

FUEL EFFICIENCY—how well a machine uses its fuel without wasting it.

FRICTION—the rubbing force between two objects sliding past each other.

HYDRAULICS—a system that pushes on a liquid with pistons to move a machine's arms and buckets.

JIB—the long arm of a crane that has a pulley on it used to lift large objects.

OUTRIGGERS—supportive legs that extend from a truck to lift its wheels off the ground and make it more stable.

PISTON—a shaft of metal that can be moved in and out of a cylinder to transmit power.

PRESSURE—a description of how much force there is acting on a surface.

PULLEY—a device that uses sets of wheels to make objects easier to lift with a cable.

ROVER—a vehicle that can travel across almost any type of surface.

SILT—the very fine earth that is carried by river currents. Silt is dropped to the bottom of large rivers where they meet the sea.

TRUSS—a series of steel or wood triangles that make structures very strong.

WIND FARM—an area of land that has many wind turbines used to make electricity.

FURTHER INFORMATION

Books to read:

Colossal Earthmovers by Keith Haddock. MBI Publishing; Osceola, WI, 2000.

Earthmovers by Hal Rogers. Child's World; Eden Prairie, MN, 1999.

Web sites to look at:

http://www.jcb.com
http://www.cat.com
http://www.bigtoy.com
http://www.bucyrus.com

A tracked excavator scoops up a load of earth.

Index

PICTURE CREDITS AGCO 8b and cover **Atlas Copco** 22cl **Corbis** 4bl Dean Conger, 10bl Raymond Gehman, 14b Graham Neden, 14 & 15t Lowell Georgia, p16, 17t Grant Smith **Sylvia Cordaiy Photo Library** 12b Chris Simpson **Ecoscene Photo Library** 16bl Tony Page **Mary Evans Picture Library** 5br **Genesis Space Photo Library** 28t, 29tl **Hulton Getty** 4 & 5t **Image Bank** 13t Pete Turner, 24 & 25t Guido Rossi **J.C.Bamford Excavators** 6 cover and contents, 7, 10 & 11t, 20b, 30 & 31b **NASA** 27tr **Stock Market** 22 & 23b Georgina Bowater, 24bl and cover Guido Rossi, **Travel Ink Photo and Feature Library** 23tr **TRH Pictures** 18b and cover, 19br, 21t, 25br, 26l, 27bl (t-top b-bottom r-right l-left c-center)